湿地光影
丛书
COLLECTION OF
WETLANDS IN IMAGES

HABITATS OF SWANS
IN CHINA

天鹅圣境

崔 林
Cui Lin
著/摄

CF▪PH
中国林业出版社
China Forestry Publishing House

图书在版编目（CIP）数据

天鹅圣境 / 崔林著、摄. -- 北京 ： 中国林业出版
社，2022.10
　（湿地光影丛书）
　ISBN 978-7-5219-1890-8

　Ⅰ．①天… Ⅱ．①崔… Ⅲ．①沼泽化地－普及读物
Ⅳ．①P931.7-49

　中国版本图书馆CIP数据核字(2022)第181873号

出 版 人：成　吉
总 策 划：成　吉　王佳会
策 　 划：杨长峰　肖　静
责任编辑：袁丽莉　肖　静
宣传营销：张　东　王思明
特约编辑：田　红
英文翻译：柴晚锁 寇佳宜 管世聪 张雅婷
装帧设计：依丹设计

出版发行：中国林业出版社（100009#北京市西城区刘海胡同7号）
http://www.forestry.gov.cn/lycb.html
电话：（010）83143577
E-mail：forestryxj@126.com
印刷：北京雅昌艺术印刷有限公司
版次：2022年10月第1版
印次：2022年10月第1次
开本：787mm × 1092mm　1/12
印张：18
字数：90千字
定价：320.00元

2014年初，借影像生物多样性调查所（IBE）与蜂鸟网合作，测试某品牌新款长焦镜头的机会，我有幸前往山东荣成的烟墩角村拍摄越冬的天鹅。经过几天的观察，我发现这些大天鹅的"寒假生活"过得简直太悠闲了。它们平时就在靠近岸边的湖面上游荡。高兴了就在附近飞一会儿，烦闷了也会和身边的邻居吵吵架。那小日子过得真是有声有色。

我这也是平生第一次这么长时间且近距离地观察野外的天鹅。看得越久，就会越觉得有意思。有时会想，它们是从哪里来的？为什么会到这里越冬？还能在国内哪些地方看到天鹅？

带着这些问题，我查阅了许多资料，结果是各种信息支离破碎，没有看到有关天鹅栖息状况比较全面客观的介绍。好歹咱也是北京师范大学生物系毕业的，基础的鸟类学、生态学知识是有的。一个念头油然生起：要不，我来做这个事儿吧。由此开启了寻找"中国天鹅湖"的旅程。

这些年，我从黄海之滨的山东荣成到西域明珠的新疆伊犁，从呼伦贝尔大草原到长江中下游的鄱阳湖，走访了数十个天鹅的栖息地。这些地方有一个共同的特点——都是比较典型的湿地生境。山东荣成的栖息地属于滨海湿地，青海贵德的栖息地属于河流湿地，江西鄱阳湖的栖息地属于湖泊湿地，新疆巴音布鲁克的栖息地属于沼泽湿地。天鹅作为大型水禽，对于这类水面宽阔的湿地十分依赖。无论是繁殖还是迁徙中的停歇以及越冬，都需要这类湿地的存在。

2019年，我在《中国国家地理》第6期杂志，发表了影像调查报告《中国有多少天鹅湖？探访三种天鹅栖息地》，比较全面地介绍了天鹅及其栖息地在中国的基本情况。但因篇幅所限，有些图片未能展示，有些故事未能讲述。

这次画册的出版，就算是对以上内容的补充。

画册的文字部分主要介绍我探访天鹅栖息地的过程，画册的图片按照"情""征""栖"三个部分展现了天鹅的基本行为特征和天鹅栖息地的环境特点。这样的内容安排是为了让读者首先能了解天鹅相关的知识，再欣赏天鹅的有趣行为和带给人们的美好感受，做到知识性与艺术性相结合。知其然，还能知其所以然。

In early 2014, I had the honor to join a project jointly sponsored by the Images Biodiversity Expedition (IBE) and the Fengniao Net for testing a newly-launched telephoto lens of a certain brand, which brought me on an unforgettable expedition to the Yandunjiao Village in Rongcheng, Shandong Province for photographing swans that were wintering there. After days of close observation, I was truly amazed to find how relaxing and carefree those swans were during their "winter breaks" — all that they needed to do was just hanging around at places close to the shoreline of the lake. They would go on a short excursion in the air over the lake if they felt like it, or start an inconsequential squabble with those standing in the vicinity when they were not feeling happy. What an agreeable life they were living!

That was the first time in my life to watch swans that live in the wild at such a close-up distance and for such a long span of time. The longer I watched, the more intrigued I became. I couldn't help but wonder: where have they come from? why are they wintering here? where else in China can swans be also found?

These questions prompted me to do some preliminary researches, which only brought me a pile of fragmented and sometimes even mutually conflicting information rather than some comprehensive, trust-worthy and in-depth introductions about this adorable bird. As a graduate of the Biology Department of Beijing Normal University who has had a systematic, if fundamental, exposure to ornithology and ecology, it suddenly occurred to me: why don't I do it myself? This unexpectedly idea set me on my long years of arduous but blissful expeditions in search of "Swan Lakes in China".

I have over the years been to dozens of habitats for swans that are dotted across China's vast territory, from the coastal city of Rongcheng in Shandong to Ili — the Shining Pearl of Western China — in Xinjiang Autonomous Region, from the vast Hulun Buir Prairie in Inner Mongolia to the Poyang Lake at the middle and lower reaches of the Yangtze River. A common point among all these habitats is that they are all located at niches composed of representative wetlands: coastal wetlands at Rongcheng, Shandong; riverine wetlands at Guide, Qinghai; lake wetlands at the Poyang Lake, Jiangxi; and swamp wetlands at Bayanbulak, Xinjiang. As the large-sized waterfowl, swans depend heavily on such wetlands that come with vast stretches of water surface. Be it for breeding, for a stopover to top-up energies and food during their migratory odysseys, or for wintering, the presence of such wetlands is essential.

In my report *How Many Swan Lakes are There in China? — Expedi-tions to Three Types of Swan Habitats* published in *Chinese National Geography* (Issue 6), 2019, I made a comparatively systematic survey about the distribution of swans and their habitats in China. Nevertheless, due to the limits of space, some beautiful photos and some interesting stories have been left out.

It's sincerely hoped that the publication of this eco-album will serve as a supplement to what had been left out there.

The texts in this book that precede the photos are mainly a documentation of my expeditions in search of swan habitats. The photos that follow afterwards are categorized into three parts, namely *Knots of Love*, *Migratory Odysseys* and *Safe Havens*, that are aimed to illustrate the characteristics in the habits and habitats of the bird at issue. The reason I have arranged the album this way is in order to give the readers an exposure to the basic knowledge about the swan before they start enjoying the visual feast presented by this elegant and adorable creature. Through the deliberate blending of artistic tastes and intellectual knowledge, it is hoped that the readers will be empowered not only with an understanding about what they are like, but also about why it is such a case.

❶ 集体照·河南三门峡
A Group Photo • Sanmenxia, Henan

冬去春来，在三门峡越冬的大天鹅们开始躁动起来。它们吵吵嚷嚷地集结成更大的群体，整编队列练习飞翔。待南风吹起，它们将飞向北方的繁殖地，抢占有利地盘来营造爱巢。相比之下，春季天鹅的迁徙时间通常较短；而秋季南迁时，因为当年出生的幼鸟体质较弱，路途中消耗的时间会比较长。

As winter wears away and spring approaches, whooper swans wintering at Sanmenxia are becoming increasingly restless. They bustle and hustle to assemble into ever larger groups for flying drills. Once the southern wind starts blowing, they will set out on their long journey to the north, where they will fight against each other to take command of advantageous spots for building their nests. The north-bound journey in spring usually takes a relatively shorter period of time, whereas the south-bound one in autumn tends to last longer, given that they have the younger chicks born during the year to take care of while they are on the way.

❷ 腾飞·河南三门峡
Testing the Wings • Sanmenxia, Henan

三门峡是大天鹅比较集中的越冬地。因为天鹅的数量多，所以拍摄的机会就多。眼前这一对儿天鹅奋力扇动着翅膀从湖面起飞。我用慢速摄影的技法，拍到了它们腾飞的瞬间。

An agreeable site for whooper swans to spend their winter, Sanmenxia is often densely populated by the birds, hence allowing more opportunities for photographers to take good pictures. The couple of swans in the photo were just about to take off from the lake surface when I pressed the button. Slow-movement shooting technique is used here.

❸ 夜寐•山东荣成
Naps at Dusk • Rongcheng, Shandong

当夜幕低垂，渔船靠岸归港，劳累了一天的人们回到温暖的家中休息，大天鹅们也弯着脖子将头钻进"羽绒大衣"，纷纷进入了梦乡。

As evening closes in and fishing boats are entering the harbor, people who have been working hard the whole day are ready to go home for a relaxing night. Whooper swans, with their necks cozily nestled within their "down overcoats", are also falling into their sweet dreams.

❹ 翱翔•河南三门峡
Soaring High • Sanmenxia, Henan

一对儿大天鹅，悄然掠过即将冰封的湖面。此情此景，它们诠释了优雅与浪漫。

A pair of whooper swan glide gracefully over the lake that is about to be covered in ice, presenting us a most-telling interpretation about the meaning of romance and elegance.

❺ 降落•山东荣成
Splashing Down • Rongcheng, Shandong

即将降落水面的大天鹅伸出了双脚。这宽大的脚蹼可以增加阻力，以减少在水中滑行的距离。

The whooper swan stretches out its paws as it is about to splash down to the lake. The wide paws are capable of exerting more friction and therefore cutting down the distance that the bird needs to cover as it glides across the water.

目 录 CONTENTS

探寻中国的 **天 鹅 湖**

EXPEDITIONS IN
SEARCH OF SWAN
LAKES IN CHINA

巴音布鲁克？
夏季看大天鹅也可以选择赛里木湖

Bayanbulak?
Sayram Lake Can Also Be an Alternative
Destination for You to Watch Whooper
Swans in Summer

新疆巴音布鲁克大草原的天鹅湖名闻天下。中秋前夕，我来到巴音布鲁克小镇，在游客中心坐着"天鹅号"专用大巴进入天鹅湖景区。原以为里面一定有"漫天飞舞"的天鹅在欢迎我们，可是走了半天连片鹅毛都没看到。好不容易来到一处铺着木栈道的小水泡子前，游客都下来活动筋骨，我看到有几只天鹅像大白鹅一样懒洋洋地漂在水面，面对熙攘的人群，它们连眼皮都不抬。

这难道就是传说中的天鹅湖？导游给大家解释，这个时期的天鹅家庭都带着幼雏躲在深深的沼泽草地里，不易看到。眼前这个水泡子里的天鹅是些"病号"，在此地接受疗养，并供游客合影。

车行驶了好久，停在一个山坡的脚下。到山上观景台，可以看到著

九曲十八弯·新疆巴音布鲁克 Nine-twists-and-eighteen-turns • Bayinbuluke, Xinjiang

名"九曲十八弯"。这里海拔近2500米，周围山地环抱，开都河蜿蜒盘转其中，大大小小的水泡子星罗棋布。黄昏时分，弯曲的河道如一条银带伸展到天际，它的尽头正好接住正在缓缓落下的太阳。眼前美景令人心潮澎湃，然而如导游所说，除非刻意去寻找，在"九曲十八弯"乃至巴音布鲁克的任何区域，想看到天鹅并不容易。

夏季的巴音布鲁克湿地是大天鹅重要的繁殖地。在1980年之前，一度有超过2万只天鹅来到巴音布鲁克繁殖和度夏，后来因人类活动干扰，大天鹅种群数量下降到约2000只。近20年间，大天鹅数量逐步回升。

除巴音布鲁克之外，大天鹅在我国北方的繁殖地理论上还有很多，不过和巴音布鲁克相似，各地大天鹅在繁殖期大多分散隐藏于植被茂盛的沼泽湿地，想目睹它们的身姿绝非易事。

为了探查我国境内的天鹅栖息地，我去过许多湿地。相比之下，新疆西部的赛里木湖是观察和拍摄大天鹅的不错去处。

赛里木湖位于博尔塔拉蒙古自治州的博乐市境内，湖面海拔约2070米，湖的西岸有季节性河流冲刷出的洪积扇以及浸润的湿地。与低海拔地区湖泊的水色不同，赛里木湖湖水青蓝，展示着高原湖泊独特的魅力。在赛里木湖繁殖的大天鹅不多，数十只，但相比于巴音布鲁克，这里更容易看到天鹅。此外，湖水与大天鹅交映，这优美的景色也更加符合人们对天鹅湖的内心期盼。

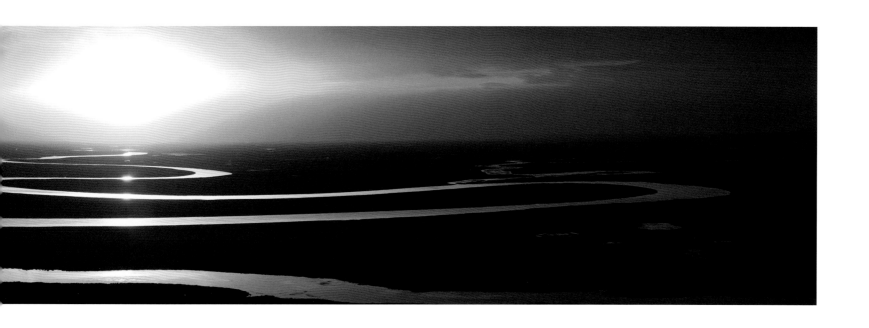

The Swan Lake in Bayanbulak Grassland in Xinjiang Province is famous all over the world. On the eve of the Mid-autumn Festival, I arrived at Bayanbulak Town and took the special bus "Swan" at the tourist center to visit the Swan Lake scenic spot. I thought there must be swans flying all over the sky to welcome us, but I didn't even see a piece of goose feather after walking for a long time. Finally, when we came to a small puddle paved with wooden walkway, tourists were down to relax their muscles, and I saw a few swans like big white geese lazily floating on the water. Facing the bustling crowd, they didn't even care to raise their eyes.

Could this be the legendary Swan Lake? The guide explained that swan families at this time were hidden with their young in deep marshy meadows, so they were not easily seen. The swans in this puddle were sick ones, recuperating here and for tourists to take pictures.

After a lengthy and tiresome drive, the bus pulled over at the foot of a hill. The famous "nine-twists-and-eighteen-turns" scenery can be seen from the observatory on the hill. Surrounded by mountains, it is nearly 2,500 meters above sea level, with the Kaidu River winding around it, dotted with puddles of different sizes. At dusk, the curved river is like a silver belt stretching all the way to the horizon where the sun is slowly setting to. The scenery is breathtaking. However, as the guide said, it is not easy to see swans in the "nine-twists-and-eighteen-turns" scenery or even in any area of Bayanbulak unless you deliberately look for them.

碧波荡漾·新疆赛里木湖　Rippling Green Lake · Sayram Lake, Xinjiang

The Bayanbulak Wetland is an important breeding ground for whooper swans in summer. Before 1980, more than 20,000 swans would come here each year to breed and spend the summer, however, the population of the whooper swans declined to about 2,000 due to the disturbance of human activities. In the past 20 years, the number of whooper swans has gradually rebounded.

Besides Bayanbulak, there are theoretically many other breeding grounds for whooper swans in northern China. However, similar to Bayanbulak, whooper swans are scattered and hidden in lush swamps and wetlands during the breeding period, so it is not easy to see them.

I have been to many wetlands to explore the habitats of swans in China.

Comparatively speaking, Sayram Lake in western Xinjiang is a good place to observe and photograph the whooper swans.

Located in Bole City of Bortala Mongol Autonomous Prefecture, Sayram Lake is about 2070 meters above sea level, with proluvial fans washed out by seasonal rivers and wetlands on the west bank. Unlike the color of lakes at low altitudes, the color of Sayram Lake is cyan, showing the unique charm of plateau lakes. Only a few dozen whooper swans are breeding in Sayram Lake, but they are easier to be seen here than in Bayanbulak. In addition, the lake and whooper swans complement each other, making this beautiful scenery more in line with people's expectations for Swan Lake.

三种野生天鹅，
不同季节营造着风格迥异的天鹅湖

Three Wild Swan Species
That Inhabit The Highly Distinctive Swan
Lakes in Different Seasons

说起天鹅湖，很多人会想到西伯利亚。遗憾的是，西伯利亚到了冬季是看不到天鹅的；而在我国，只要找对地方，任何季节都可以看到野生天鹅。探寻中国的天鹅湖，用影像展示我国的野生天鹅以及它们的栖息地，是我多年来的夙愿。

天鹅是严格依照气候周期性变化而往返迁徙的候鸟。不同的季节，天鹅会出现在不同的地域。它们繁殖和抚育幼鸟的地区叫作"繁殖地"；避寒过冬的地区叫作"越冬地"；迁徙途中休息的地区则被称作"停歇地"。

除去时间因素，每处天鹅湖可以看到的天鹅种类也不尽相同。世界上的天鹅有七种，分别是：大天鹅，小天鹅，疣鼻天鹅，黑天鹅，黑颈天鹅，黑嘴天鹅及扁嘴天鹅。在我国能见到的野生天鹅有三种，分别是：大天鹅、小天鹅和疣鼻天鹅。

三种天鹅的形态远看非常相似，都有白色的羽毛，长长的脖颈，黑色的脚趾和蹼。如若细看，它们之间的区别也很明显：大天鹅嘴部长且黄色部分较多，一直延伸过鼻孔；小天鹅的体形较大天鹅小，嘴部短且后端黄色的部分不超过鼻孔；疣鼻天鹅体形较大天鹅大，嘴部颜色橘红，前额上有一块黑色的瘤疣突起。在水面上时，大天鹅和小天鹅的脖颈相对都比较直，疣鼻天鹅的脖颈则通常呈"S"形。

在食物上，三种天鹅吃的都比较素，以水生植物的根、茎、叶和种子为主，偶尔也吃小的软体动物及昆虫。因此，水草丰美的湖泊、河流、海滨及库塘等湿地环境是它们最惬意的生境。

Speaking of Swan Lakes, many people would think of Siberia. Unfortunately, you won't be able to see swans in Siberia in winter; while in China you can see wild swans in any season as long as you find the right place. It has been my long-cherished wish for many years to explore the Swan Lakes in China and document the wild swans and their habitats through visual images.

Swans are migrant birds that migrate in strict accordance with the cyclical climate change. Swans can be found in different areas in different seasons. The area where they breed and raise their young is called the "breeding ground"; the area where they shelter from the cold and spend the winter is called the "wintering ground", and the area where they rest during migration is called the "stopover".

Apart from the time factor, the species of swans that can be seen in each Swan Lake are also different. There are seven species of swans in the world, namely: whooper swan (*Cygnus cygnus*), tundra swan (*Cygnus columbianus*), mute swan (*Cygnus olor*), black swan (*Cygnus atratus*), black-necked swan (*Cygnus melancoryphus*), trumpeter swan (*Cygnus buccinator*) and coscoroba swan (*Coscoroba coscoroba*). In China, we have three species of swans: whooper swan, tundra swan and mute swan.

These three species of swans look very similar from a distance, with white feathers, long necks, black toes and web. But when you look closely, the difference between them is also obvious: the bill of the whooper swan is long, with more yellow patches extending all the way through the nostrils. The tundra swan is smaller than whooper swan, with a short bill and a yellow patchon the base of its bill that does not exceed the nostrils. The mute swan is larger than the whooper swan, with a reddish-orange bill and a black knob on the forehead. On the water, the necks of both whooper swan and tundra swan are relatively straight, while the necks of the mute swan are S-shaped.

In terms of food, these three species of swans are vegetarians, mainly feeding on the roots, stems, leaves and seeds of aquatic plants, and occasionally also on small molluscs and insects. Therefore, wetlands such as lakes, rivers, seashores, reservoirs and ponds with abundant water and grass are their most favorable habitats.

天鹅 whooper swan (*Cygnus cygnus*)　小天鹅 tundra swan (*Cygnus columbianus*)　疣鼻天鹅 mute swan (*Cygnus olor*)

大天鹅寒冬集中在山东荣成、河南三门峡和山西平陆

Whooper Swans Mainly Gather in Rongcheng City in Shandong Province, Sanmenxia City in Henan Province and Pinglu County in Shanxi Province

山东荣成是大天鹅聚集越冬的著名栖息地。荣成位于山东半岛的最东端，沿海浅滩拥有许多淡水注入的潟湖，潟湖岸边生长着成片的芦苇丛，夏季在蒙古国北部繁殖的大天鹅种群长期来此地越冬。荣成的沿海盛产大天鹅爱吃的水生维管植物——大叶藻，另外，周边农田中的冬小麦青苗也是大天鹅喜爱的零食。荣成能看到天鹅的地点有：烟墩角、天鹅湖景区、马山、车祝沟、绿岛湖等。每年蒙古国北部的大天鹅来此越

滨海湿地·山东荣成　Coastal Wetland·Rongcheng, Shandong

冬，总数能有数千只。

我国境内大天鹅越冬的湿地远不止山东荣成这一处。在黄河的三门峡段两岸，河南省三门峡市与山西省平陆县的湖泊湿地同样是大天鹅非常重要的越冬地。

虽说从行政区划来看，三门峡市与平陆县分属两省，但它们同属于黄河中游区段范围，位于三门峡库区两岸。在这片湖泊湿地越冬的大天鹅也来自蒙古国北部，它们途经内蒙古和陕西，每年10月下旬到达三门峡一带，直到次年3月上旬集群北飞。

根据以往数据来估测，冬季在我国境内越冬的大天鹅种群总数量为1万~1.5万只。而如今在三门峡段黄河湿地越冬的大天鹅总数可能接近1万只，这里已然成为国内大天鹅冬季数量最多的栖息地。

Rongcheng in Shandong is a famous wintering habitat for whooper swans. Located at the easternmost tip of Shandong Peninsula, Rongcheng has many freshwater-fed lagoons in the coastal shoals, and the lagoon shores are lined with patches of reeds. The whooper swans that breed in northern Mongolia in summer have long established a habit to come here to spend the winter. The coastal areas of Rongcheng are rich in eelgrass, an aquatic vascular plant that whooper swans love to eat. In addition, winter wheat seedlings in the surrounding farmland are also favorite "snacks" for them. Whooper swans can be found in these places: Yandunjiao, Swan Lake Scenic Area, Mashan, Chezhugou, Green Island Lake, *etc*. Every year, thousands of whooper swans from northern Mongolia would come to spend the winter here.

The wintering wetlands for the whooper swan in China are by no means limited only to the one in Rongcheng, Shandong. On both sides of the Sanmenxia section of the Yellow River, the lakes and wetlands of Sanmenxia City in Henan Province and Pinglu County in Shanxi Province are

大鹅湖景区•河南三门峡　Swan Lake • Sanmenxia, Henan

also very important wintering grounds for whooper swans.

Although Sanmenxia City and Pinglu County belong to two provinces from the perspective of administrative division, they are both located in the middle reaches of the Yellow River, sitting on either side of the Sanmenxia Reservoir Area. Whooper swans that overwinter here also come from northern Mongolia. They pass through Inner Mongolia Autonomous Region and Shaanxi Province, arriving at Sanmenxia City in late October every year and staying here until early March of the next year when they start to fly northwards.

According to previous data, it is estimated that the total population of whooper swans wintering in China is about 10,000 to 15,000. Now, the total number of whooper swans wintering in Sanmenxia section of the Yellow River wetland stands close to 10,000, making this place the habitat that has the largest whooper swan population in China during winters.

雪霁·山西平陆　A Sunny Day after Snow · Pinglu, Shanxi

青海与新疆，
大天鹅在寒冷北方亦可越冬

Qinghai and Xinjiang Are Two Places
in The Cold Northern Part of China
Where Whooper Swans Can Also Spend Their Winter Times

此外，在我国西北部也有大天鹅的越冬地。青海湖的冬季风大寒冷，夜晚气温可以达到零下30℃。但湖西岸石乃亥乡的尕日拉寺附近的湿地，大批的大天鹅在这里栖息聚集，它们主要来自新疆巴音郭楞蒙古自治州的巴音布鲁克。这里泉水分布，冬季不易结冰。看来如果有流动的活水和充沛的食物，哪怕狂风与严寒，天鹅也可以容忍。

站在岸边拍摄，厚实的鞋袜并不能让脚感到温暖，我十个脚趾被冻到没有知觉，但天鹅的脚却并不会冻伤。它们的体温约38℃，但脚趾的表面温度仅仅比冰点稍高。因为脚内的细胞液含量稀少，血液循环速度很快，而且血液流入和流出双脚的血管彼此相邻，在"逆流热交换效应"下，动脉血的热量可以转移到静脉血中，继而返回体内。这样，只需要消耗很少的热量就可以避免脚被冻伤。

同在青海东部的贵德，冬季也有大天鹅栖息。这里地理条件优越，素有"高原小江南"的美誉。清澈的黄河水从城北缓缓流过，河滩两岸植被茂盛。几十只大天鹅自在地漂荡水中，风景如画。再往西，新疆也有多处大天鹅的越冬地，比如，博乐市以及库尔勒市的市区。发源于博斯腾湖的孔雀河从库尔勒市穿城而过，冬天虽冷，但河水长流不易结冰，几十只大天鹅在孔雀河中越冬。

In addition, there are also areas in northwestern China for whooper swans to get through the cold winters. In winter, the Qinghai Lake is featured with strong wind and low temperature which sometimes might drop as low as minus 30 degree Celsius during late nights. However, over the vast wetlands in the Ga'rila Temple in Shinaihai Township that sits on the west bank of the lake, large numbers of whooper swans would arrive from Bayanbulak of Bayingol Mongolian Autonomous Prefecture in Xinjiang for wintering. Springs that do not freeze in winter are densely distributed here.

It seems that, so long as there is running water and sufficient food supply available in a given place, it will be likely to become a tolerable winter habitat for swans despite of the severe storms and gushing winds that may potentially strike.

In spite of thick socks and shoes I wear, I hardly feel any warmth in my feet and the ten toes are almost numb as when I stand on the bank to take photos. Nevertheless, the feet of the swans won't be harmed by frostbites. The body temperature of swans is about 38 degree Celsius, but that at the surface of their toes is just slightly above the freezing point. The reasons are as follows: there is only scant cellsap in their feet and the pace of blood circulation is fast. Besides this, the blood vessels that are responsible for transporting blood into and out of their feet are adjacent to each other. Under the influence of the "countercurrent heat exchange effects", heat in arterial blood can be firstly transmitted into venous blood and then into

their bodies. As a result, they are able to keep their feet safely away from frostbites with very little energy consumption.

Whooper swans are also found to exist in winters in Guide County, which is likewise situated in eastern Qinghai. For the presence of favorable geographical and natural conditions, this county is acclaimed as "the place on the plateau that comes with Jiangnan-like scenery". The crystal-clear Yellow River flows quietly by on the northern part of the city, with either bank of which densely covered by lush vegetations. Dozens of whooper swans swim leisurely in water, among the picturesque view. Wintering sites for whooper swans are also available further westward in Xinjiang, with the urban areas of Bole and Korla as typical examples. The Bosten Lake-originated Kongque River flows quietly through the center of Korla City, where dozens of whooper swan can often be found wintering in the freeze-free water.

湖岸湿地•青海青海湖　Lakeshore Wetland • Qinghai Lake, Qinghai

小天鹅越冬地
集中在长江中下游湿地

Wintering Grounds for Tundra Swan:
Concentrated in Yangtze Plain Wetlands

　　小天鹅不是小时候的大天鹅，而是完全不同于大天鹅的另一个物种。小天鹅的繁殖地比大天鹅的繁殖地更加靠北，主要位于欧亚大陆北部北极圈内的苔原地带，东西横贯俄罗斯北冰洋沿岸大部分低地和平原。我国境内没有小天鹅的繁殖地，理论上夏季在国内见不到野生的小天鹅。但是到了秋冬季，小天鹅会集中在我国长江中下游平原的湖泊与湿地。因为这个季节的鄱阳湖、洞庭湖、太湖以及各个大小湖泊水位普遍降低，湖边会形成大面积的浅滩湿地。充沛的食物资源加上优越的气

栖息的小天鹅•江西九江　Roosting Tundra Swans • Jiujiang, Jiangxi

候条件，吸引了包括小天鹅在内的众多鸟类来此过冬。

在江西九江的东湖生机林、都昌的马影湖、鄱阳的白沙洲、余干的康山大堤和吴城的大湖池等湖边湿地，冬季都可以看到大群的小天鹅。安徽池州的升金湖、江苏苏州的太湖、上海浦东南汇的东滩以及崇明岛也是观察小天鹅的好去处，只是小天鹅警觉性比较高，不太容易靠近。

小天鹅每年春季2~3月开始迁徙，待到5月下旬飞抵北极圈内的苔原地带繁殖，迁徙距离长达5000~7500千米。因为路途遥远，它们途中停歇时间也较长。根据给在鄱阳湖越冬的小天鹅安装的卫星定位装置返回的信息，它们会飞向位于东西伯利亚和西西伯利亚的不同的繁殖地。因为迁徙路径并不单一，故而在小天鹅的迁徙季节，我国长江以北的很多地区都有机会目睹它们的身影。

飞过海滩•上海浦东　Flying over the Beach • Pudong, Shanghai

Tundra swans, whose name in Chinese literally goes like "the little swan", do not refer to the younger chicks of whooper swans, whose name in Chinese literally goes like "the large swan". Instead, the two of them belong to two distinctive species of swans. Compared with that of whooper swans, the breeding grounds of tundra swans tend to be situated further northwards, mainly in the tundra zone of northern Eurasia continent that falls within the Arctic circle, as well as in a major part of the low-lying plains along the coast of the Arctic that spans Russia in east-west direction. No breeding grounds for tundra swans have so far been found in China, so theoretically speaking, it is not possible for people to see new-born wild-inhabiting tundra swans in China during the summer season. However, as autumn and winter set in, they will be found in massive flocks over the vast lakes and wetlands across the Yangtze Plain, where the water level of such major lakes as the Poyang Lake, the Dongting Lake and Lake Taihu will on the whole begin to fall, leading in result to the formation of extensive stretches of fords and wetlands on the outskirts of these lakes. The mild climate here, plus abundant food supply, makes this region a most favorable niche that attracts tundra swans and other waterfowls to come here to spend their winters.

In Jiangxi, large flocks of tundra swans can be found in East Lake of Jiu-

集群的小天鹅·江西鄱阳湖　AssemblingTundra Swans · Poyang Lake, Jiangxi

jiang, Majing Lake of Duchang, Baishazhou Nature Reserve of Poyang, Kang-shan Great Dyke of Yugan, Lake Dahuchi of Wucheng and other lake-side wetlands in winter. In addition to the previous places, Shengjin Lake of Chi-zhou in Anhui, Lake Taihu of Suzhou in Jiangsu, Nanhui Dongtai Wetland of Pudong New Area and the Chongming Island in Shanghai are also the favored destinations for people interested in having a close watch of tundra swans. The only thing you need to take into consideration is that the bird is highly vigilant by nature and therefore hard to get close to.

Tundra swans would typically set out on their north-bound migratory journeys at sometimes between February and March in spring, covering a to-tal distance of 5,000 to 7,500 kilometers before they reach in the tundra zone within the Arctic circle to breed in late May. Due to the long distance, they will need to stop for a long period of time to rest during the journeys. Accord-ing to the feedback derived from satellite positioning devices tied to the tundra swans wintering in Poyang Lake, they tend to fly to different breeding grounds in eastern and western Siberia. Because the migration routines they follow vary significantly, it is likely for people to see them in many regions in China that sit to the north of the Yangtze River throughout the migratory seasons.

巴彦淖尔与呼伦贝尔，
隐藏着疣鼻天鹅的繁殖地

Bayannur and Hulun Buir,
the Hidden Breeding Grounds for Mute Swans

在欧洲，疣鼻天鹅很常见，有些就生活在城市中的湿地公园或河道里。但在我国，疣鼻天鹅就比较稀罕了。它们在国内的繁殖区域，主要集中在内蒙古自治区的几处湿地。

从卫星地图上看，内蒙古乌拉特前旗北部有一片"肾形"的绿洲，那就是被称为"塞外明珠"的乌梁素海。根据近几年数据估测，每年来乌梁素海繁殖的疣鼻天鹅数量为500~600只。

乌梁素海东岸靠北的区域芦苇比较稠密。这里便是疣鼻天鹅春夏季节求偶与繁衍后代的场所。疣鼻天鹅恋爱的仪式感很强，可以经常看到成对儿的天鹅弯曲脖颈面对面"秀爱心"；有时也可以看到一对儿天鹅同时向左或者向右转头，像在跳慢节奏的探戈。

在内蒙古东部的呼伦贝尔大草原，呼伦湖与贝尔湖之间有一条狭长的乌尔逊河。乌尔逊河的河道旁有一处名为乌兰泡的牛轭湖，这里人迹罕至、芦苇密集。每年春夏时节，众多的候鸟回来营巢繁育后代，其中也有疣鼻天鹅，其种群数量约100只。

The mute swan is a species commonly found in wetland parks and rivers in cities across Europe. However, this species of swan is rarely seen in China in comparison, since their breeding areas in the country are limited only to several wetlands located in the Inner Mongolia Autonomous Region.

From the satellite map, an oasis which looks like a kidney can be seen in northern Urad Front Banner of the Inner Mongolia Autonomous Region. The oasis is the Ulansuhai Nur, known as "the pearl beyond the Great Wall". Based on statistics obtained in recent years, it is estimated that the number of mute swans breeding in the Ulansuhai Nur stands roughly around 500 – 600.

The northern part of the east bank of the Ulansuhai Nur is covered with dense groves of thriving reeds, making it a safe haven for mute swans to seek mates and give birth to their offspring during the spring and summer seasons. Mute swans attach great importance to the sense of ritual so far as love is concerned, thus they are often seen to bend their necks in face-to-face positions to put on a sign of love that resembles the shape of a heart. Sometimes, they can also be found to turn their heads left or right in synchronization, as if they are tangoing slowly.

On the Hulun Buir Grassland in eastern Inner Mongolia, a narrow but long river, the Ursun River, meanders between the Hulun Lake and the Buir Lake. Near the course of the Ursun River is an out-of-way oxbow-shaped lake — the Wulanpao Lake — that is densely covered with thriving reeds. In spring and summer every year, many migratory birds would arrive here to give birth to their offspring, included among which would be mute swans, whose population stands approximately around 100.

呵护•内蒙古乌兰淖尔　Caring Kinship • Ulan Nur, Inner Mongolia

巴尔喀什湖的疣鼻天鹅
冬季旅居伊犁河谷"天鹅泉"

Mute Swans Born in Lake Balkhash
Taking a Winter Break at "Tian'equa (The Swan Spring)"
in The Ili Valley

到了冬季，在内蒙古等地繁殖的疣鼻天鹅，会迁徙到朝鲜半岛一带。而夏季栖息于中亚的一群疣鼻天鹅，其中一部分悄然来到我国新疆西部的伊犁河谷越冬。

伊犁河谷的冬季十分寒冷，但是位于伊宁县英塔木镇的"天鹅泉"异常火爆。全国各地的摄影爱好者纷至沓来，力争拍到天鹅大片。原来这里有泉水，冬季水塘里不结冰，每年深秋会有数百只疣鼻天鹅栖居于此。如果赶上有雾凇的天气，岸边树木银装素裹，池塘里雾气缥缈，几只洁白的天鹅游荡其中，场景如梦如幻。

这批疣鼻天鹅来自西边的哈萨克斯坦，巴尔喀什湖是它们的繁殖地。"天鹅泉"不过是片看似普通的水塘，据水塘的主人韩新林讲，他家一直利用泉水养鱼，也养些土鹅、土鸭。1993年，几只疣鼻天鹅第一次来到他家鱼塘过冬，韩新林热情款待。后来，每年来的天鹅越来越多，他就与家里的鸭鹅混养。开饭的时候，他喊着"鸭鸭鸭"招呼它们一起用餐。在此越冬期间，疣鼻天鹅不愁吃喝，不受惊扰，逍遥自在。

为了照顾好疣鼻天鹅，韩新林和儿子韩亮丝毫不敢怠慢。他们干脆把鱼塘改造，将水面加长，以适合天鹅的滑翔起降。近些年，冬季来他家的天鹅已经有几百只。这个小小的"天鹅泉"已经成为著名的天鹅摄影基地。

In winter, the mute swans that breed in areas such as Inner Mongolia will migrate to Korean Peninsula, while some of those which inhabit Central Asia in summer will migrate to the Ili Valley in western Xinjiang to spend their winter times.

Winters in the Ili Valley tend to be extremely cold, but the "Tian'equan" located in Yingtamu Town in Yining County is very popular. Avid photographers from all over the country flock there, in eager anticipation of shooting their stunning photos of swans. Thanks to the presence of freeze-free spring here, hundreds of mute swans would arrive here in late autumn to take it as their wintering sites. If viewers are fortunate enough to arrive on days with rime, they would have the chance to appreciate the beautiful scenery that only can be seen in dreamlands — the snow-clad trees, the misty ponds, the snow-mute swans that swim leisurely··· It's just incredible!

Born in Lake Balkhash, these mute swans come from Kazakhstan that sits to the west of China and they breed in the. "Tian'equan" apparently looks just like an ordinary pond that, according to Han Xinlin, the proprietor of this pond, has been used for many years to raises fish, geese and ducks. Back in 1993, several mute swans arrived for the first time at his pond to winter there, who were treated with great hospitality by Mr. Han. The swans have come regularly ever since, and in ever increasingly populations. So, he started to raise them together with his domestically-raised ducks and geese. "Duck, duck, duck", he would shout to assemble them in to eat at dinner times. And so, the mute swans start to spend their carefree winters here, well-fed and well-taken care of, with nothing to worry about.

In order to better feed the mute swans, Han Xinlin and his son Han Liangsi care them without even a second of neglect. They even converted the pond to a longer one that adapts better to the needs of the swans in taking off and splashing down. Over recent years, the number of swans wintering in his pond has risen to hundreds, making this little "Tian'equan" a renowned destination for swan-loving photographers across the country.

伊犁河畔·新疆伊犁　Ili River · Ili, Xinjiang

天仙下凡·新疆伊犁　Celestial Fairies in Human World · Ili, Xinjiang

不同时节，
我们该去哪里欣赏天鹅

Where Should We Go to See Swans in
Different Seasons?

有朋友可能会说，在他们家附近的湿地公园或野生动物园还能看到黑天鹅，但这种黑天鹅是大洋洲引进物种，其实已经家禽化了。它们在人工环境下安逸地生活，没有迁徙的特性。

经过几年在国内的实地走访和追寻，我对三种野生天鹅的基本情况作个小结。冬季，看大天鹅可以去河南三门峡、山西平陆、山东荣成以及辽宁朝阳，看小天鹅可以去长江中下游湿地，想看疣鼻天鹅那就只有去伊犁河谷了。夏季，看大天鹅可以去新疆的赛里木湖，看疣鼻天鹅可以去内蒙古的乌梁素海，小天鹅应该是看不到的。特别需要注意的是，春夏是天鹅的繁殖季，它们都躲在隐蔽的水草湿地，我们千万不可过于

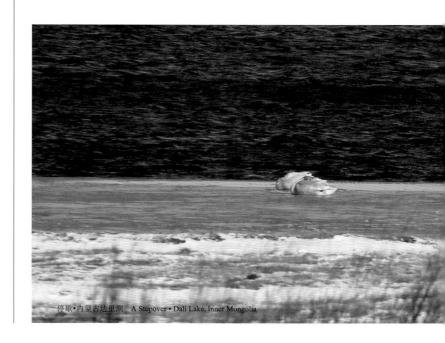

停歇·内蒙古达里湖   A Stopover · Dali Lake, Inner Mongolia

靠近，避免惊扰它们营巢或育雏。

　　春季和秋季是候鸟的迁徙季，我国的西北和西南地区，处在"东非-西亚迁徙路线"和"中亚-印度迁徙路线"上；我国的东北、华北、华中、华东及华南地区处在"东亚-澳大利西亚迁徙路线"上。这些地区的河流、湖泊、沼泽及滨海湿地是候鸟的"服务区"和"加油站"。除了华南地区，在较宽阔的湿地环境都有可能看到天鹅。只是它们的"旅行团队"有大有小、有早有迟，停留的时间不固定。能否见到需要靠运气。

　　深秋，内蒙古克什克腾旗的达里湖有些地方已经结冰，当我站在西南的曼陀山上仍然可以看到湖边湿地有许多大天鹅在此休整集结。在河

北沽源的囫囵淖尔的天鹅湖可以看到成群的小天鹅徜徉其中。在辽宁朝阳的白石水库湿地能见到大天鹅翩翩起舞。天津的北大港湿地、山东东营的黄河口湿地也都是天鹅必经的停歇地。北京的官厅水库野鸭湖、怀柔水库、密云水库附近都有天鹅活动，就连在颐和园、圆明园都有可能看到天鹅的身影。

　　伴随着我国自然生态环境的不断改善，在信息传播迅猛的当下，我相信有关天鹅的栖息地，今后必然会有更多新的发现。

Some people might say that they can still see black swans in the wetland parks or wildlife parks near their home. But the fact is that these kinds of swans belong to a species introduced from Australia and have already been domesticated. They live comfortably in artificial environments and therefore have lost their migratory habits.

Drawing on the field observations and researches that I have personally made over the past years at swan habitats situated across China, I have reached the following general conclusions concerning the rough profile of the three species of wild swans. For the winter months, if you want to watch whooper swans, you should go to Sanmenxia in Henan, Pinglu in Shanxi, Rongcheng in Shandong and Chaoyang in Liaoning. To watch tundra swans, you can go to the wetlands in the middle and lower reaches of the Yangtze River. And if your favored bird is mute swans, the only domestic option for you would be the Ili River Valley. As to the summer, Sayram Lake in Xinjiang is a good destination for watching whooper swans, and Ulansuhai Nur in Inner Mongolia for mute swans, whereas it's unlikely for us to come across the tundra swans anywhere across the country. What needs to be pointed out is that spring and summer are the breeding seasons for swans, when swans tend

集结·天津北大港　Assembling for the Journey · Beidagang, Tianjin

46

to be hidden in safe and hard-to-access shelters in densely covered wetlands. It's highly unadvisable for people to get too close to them in case they be disturbed and therefore unable go ahead smoothly with their nest-building or brooding activities.

Spring and autumn are the seasons when migratory birds would be on their long odysseys. The northwestern and southwestern regions of China are on the "East Africa-West Asia Migration Route" and "Central Asia-India Migration Route", while the "East Asia-Australia Migration Route" covers China's northeastern, northern, central, eastern and southern regions. Rivers, lakes, marshes and coastal wetlands dotted across these regions are often the "service areas" or "gas stations" for the migratory birds to top-up energy supplier or to relieve themselves of fatigues. Except for certain parts in South China, swans can be found in almost all other regions across the country so long as there are vast stretches of wetlands. The only difference is that the size of their communities can be large or small, the specific time for them to arrive may vary more or less, and there is no certainty as to how long they tend to stay in a given place. Whether or not you can see them is a sheer matter of luck.

闲庭信步·北京密云　Leisurely Stroll · Miyun, Beijing

In late autumn, some parts of Dali Lake in Hexigten Banner, Inner Mongolia have frozen over. As I stand on the top of the Mantuo Mountain that stands toweringly to the southwest of the lake, I can still see large flocks of whooper swans assembled in the wetlands scattered around the lake. You can see flocks of tundra swans roaming leisurely in the Swan Lake in Hulun Nou'er in Guyuan, Hebei, and massive of whooper swans dancing their graceful dances onthe wetlands around the Baishi Reservoir in Chaoyang, Liaoning. The Beidagang Wetland in Tianjin and the Yellow River Delta Wetland in Dongying, Shandong are also the critical stopover places for swans on their migratory journeys. In Beijing, swans can be found near the Wild Duck Lake of Guanting Reservoir, Huairou Reservoir and Miyun Reservoir. Even in the urban-situated royal parks like the Summer Palace and Yuanmingyuan Imperial Garden, it is not rare for visitors to come across their graceful figures.

Along with the continuous improvement in natural ecological environment in China and the rapid information spreading of the present age, I believe increasingly more new discoveries concerning swan habitats will surely be made in the future.

掠过湖面·江苏太湖 Splashing across the Lake • Taihu Lake, Jiangsu

HABITATS OF SWANS
IN CHINA

天鹅圣境

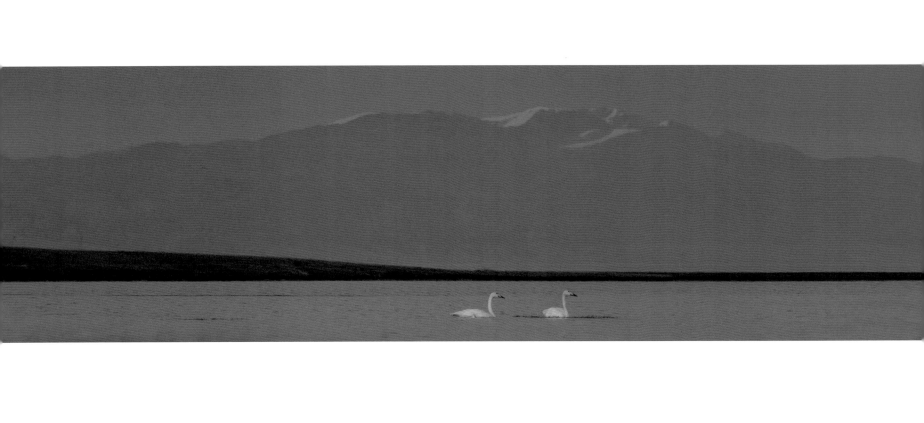

情

天鹅绝对是有情感的动物。经过我持久地观察，发现它们也有喜怒哀乐的情感表达。

处在恋爱中的天鹅会表现出明显的不寻常的行为，它们会彼此靠近并共同做出一致的动作，就像一对舞伴在跳华尔兹或探戈。这种"舞蹈"可以理解为是培养彼此动作的协调性，是对未来一起生活能否达成默契的锻炼。

成了对儿的天鹅会终身相守，无论干什么都是形影不离。若是发现心怀不善者靠近，一般雄性会展现出攻击姿态驱离对方。这种肢体动作加上表情，你能充分感受到它的威慑力。在育雏期的天鹅夫妇则更加敏感，除了驱逐还会用嘴撕咬对方，并且是穷追猛打，非常凶狠。

在繁殖地，天鹅一般以家庭为单位，过着自己的小日子。但是在迁徙的过程中，出于安全的需要，它们会集结成群。到了越冬地，更是要融入大集体之中。在这样相对拥挤的环境中，它们彼此学会了迁就包容，呈现出一派祥和的局面，当然，偶尔也会发生争执，甚至大打出手。

当我看懂了天鹅的这些爱恨情仇，不由地会心一笑：人类亦是如此。

Swans are definitely a sort of emotional creature. Close observations convince me that swans have their unique ways for expressing their inner feelings, be the feeling related with happiness or sorrow.

Swans in love tend to show some special signs that are distinctively different from their normal behavior. They would intentionally stay close to each other and synchronize in their actions, just like two human beings waltzing or tangoing together. Their "dances" can be interpreted as a training that is aimed to foster a tacit consensus in their movements as well as in their attitudes towards future life.

Swans, once the knots of love are tied, will stay loyal to each other all through their lives, keeping close company whatever they do. Once it detects any slight ill intentions in anyone approaching its partner, the male swan will immediately become aggressive and belligerent, as if trying to scare the intruder away. You can obviously feel the deterring power in its physical actions and facial expression. The couples in their brooding stage would often be even more sensitive and assertive, often pecking the intruder fiercely with their tough mouths until the latter acknowledges defeat and flee away.

Swans typically live their independent lives in units of family on their breeding grounds. Nevertheless, for the sake of security, they tend to assemble into large flocks during their migratory journeys. When they arrive at their wintering grounds, they are likely to get into still larger and more crowded communities, in which they appear to be mutually understanding and accommodating — the community is steeped in a peaceful and harmonious mood. For sure, disputes do break out occasionally, leading sometimes even into physical violence.

Now that I've got a better understanding about the emotional world of swans, I can't help but break into a knowingly chuckle: isn't it the same case with us human beings?

● "双人舞" · 新疆伊犁
Pas de deux • Ili, Xinjiang

　　恋爱中的疣鼻天鹅经常会一起跳"双人舞"。有时彼此转圈，跳的是"华尔兹"；有时左右甩头，跳的是"探戈"；有时面面相对，胸部贴在一起，这跳的算是"贴面舞"吧。

　　Mute swans in love would often stage *pas de deux*. Sometimes they would spin and twist around each other in waltz; sometimes they would flip their heads in tango; and sometimes they would dance face to face, as if engaged in an intimate *pas de deux* that belong only to the two of them.

▶ 示爱·河南省三门峡
Pledge of Love • Sanmenxia, Henan

　　大天鹅表达爱情的方式相对于疣鼻天鹅就比较简单了。通常雄性天鹅在雌性面前起身振翅引吭高歌，雌性天鹅则仰面应和，崇拜之情溢于言表。

　　Compared with mute swans, the way whooper swans pledge their love is much simpler. Usually, the male one usually would just, with his body straightened up and wings flapping vigorously, sing his melodious love songs loudly, whereas the female one would reciprocate her loving melodies, face beaming with admiration and adoration.

◀ 警惕·山东荣成
Vigilance • Rongcheng, Shandong

　　一对大天鹅正在缓缓地相互靠近，身体轻柔地触碰，彼此都沉浸在柔情蜜意中。我被这迷人的时刻吸引，试图更靠近一些，结果被察觉到了。它们向我投来警惕的目光。

　　A couple of whooper swans were swimming increasingly closer to each other. Their bodies were gently touching and they were both bathed in tenderness. Intrigued by this touching scene, I tried to get even closer without disturbing them but fail. They caught sight of me and cast a vigilant look at me.

驱赶·新疆伊犁
A Showdown • Ili, Xinjiang

　　恋爱中的雄性疣鼻天鹅，对潜在的竞争对手会表现出很强的攻击性。有时会乍起翅膀、脖子向后冲向对方。整个身体像拉满的弓，嘴像是即将离弦的箭。那咄咄逼人的气势令对手胆寒。

　　A male mute swan that is in love tends to be extremely belligerent and aggressive against potential competitors. With his wings erected angrily and neck bent backwards, he will sometimes thrust his body forcefully towards his rival. If we compare his body to a fully stretched bow, his tough mouth would be like an arrow that is about to be let go. The assertive air he puts on is prohibitive enough to drive any ill intention bearing competitor away.

◀ 打击·新疆伊犁
Love-stricken • Ili, Xinjiang

　　一只单身雄性疣鼻天鹅靠近了一对儿正在恋爱中的情侣，引起了它们的强烈不满，结果遭到了无情地打击。

　　A male single mute swan is intruding upon the territory of a couple that is in ferment love, causing strong protests from the latter, who jointly force the intruder to flee in a disheartened state.

憧憬·山东荣成
Longing • Rongcheng, Shandong

　　阳光下的水面波光粼粼，在镜头中呈现出梦幻的光斑。一对儿大天鹅背朝着我缓缓地游向远方，不知此刻它们在想些什么，也许在憧憬着美好的未来。

　　The sparkling surface of lake under the sunshine projects beautiful light spots on to my camera lens. A couple of whooper swans are leisurely swimming into the distance, with their backs turned to me. What they are thinking about at this moment, I wonder, engaged perhaps in a sort of longing for a better future?

● 私语·新疆赛里木湖

Whisper • The Sayram Lake, Xinjiang

初秋时节的，新疆赛里木湖畔雪峰初现，湖面一派宁静安详。一对大天鹅在窃窃私语，也许正在商量着入冬前的远行。

In early autumn, snow-capped peaks begin to show their initial appearance around the tranquil and serene Sayram Lake located in Xinjiang. A couple of whooper swans are whispering to the ears of each other, perhaps deliberating on the upcoming long journey they will need to take before winter sets in.

平湖秋月·新疆赛里木湖
Autumn Moon over the Calm Lake • The Sayram Lake, Xinjiang

农历八月十五的傍晚，我驱车来到赛里木湖。当时景区还在修建中，不能提供住宿，我只能在车里蜷缩一宿，所以天不亮就来到了湖边。可能是上苍感受到了我的诚意，安排了一对儿天鹅在不远处静静地等候。好一幅天鹅版的"平湖秋月"。这眼前的一幕，是我不曾想到的，但又是真真切切亲眼所见。冥冥之中，最美天鹅湖，居然就这样展现在了我的眼前。

At the night of the fifth day of the eighth lunar month, I drove to the Sayram Lake. As the scenic spot was undergoing construction and unable to provide accommodations, the only thing I could do was to put myself up in the car for the night. But as luck would have it, this unlucky event made it possible for me to arrive at the lake before dawn. It must be my sincerity that has moved the God, who dispatched a couple of swans there to wait for me patiently. What a nice view of "Autumn Moon Over the Calm Lake" complemented by the presence of the swans! What lay in front of me was something that I had never imagined, but it was indeed there, so vivid and real, right in front of my eyes. As if pre-destined, I were able to look at the most beautiful swan lake in the country under such an unexpected circumstance.

觅食·天津北大港
Foraging • Beidagang, Tianjin

一对小天鹅将脖子深深地探入水下，只露出尾巴。它们不是在做游戏，而是在水下寻找食物。

A couple of tundra swans are dipping their neck deep into the water, leaving only their tails visible to people standing on the banks of the lake. Instead of being engaged in a certain game, they are in fact foraging for food under the water.

⊙ 比翼双飞·辽宁朝阳
Flying Wing to Wing • Chaoyang, Liaoning

　　已经成了对儿的天鹅，无论干什么都是要在一起的。双宿双飞，形影不离。

　　No matter what they do, the swans that have tied their love knots will be doing it together, never parting for even a single moment.

◀ 隐现·内蒙古乌梁素海
The Dimly-visible View • Ulansuhai Nur, Inner Mongolia

　　一对疣鼻天鹅游向芦苇荡的深处。透过眼前枝叶的缝隙，还能看到它们时隐时现的身影。为了安全，它们的巢区通常比较隐秘。

　　A couple of mute swans are swimming towards the depth of reeds. Through the leaves, I can vaguely see their figures. For security reasons, their nests are often built at places where it is relatively more difficult to detect. 75

◔ 巡游·内蒙古呼伦贝尔
Escorted Cruise • Hulun Buir, Inner Mongolia

　　一只苍鹭站在茂密的芦苇荡里等着鱼儿的靠近。这时，一对疣鼻天鹅夫妇带着六只幼崽静静地游过它的身边。队伍的最前面是天鹅妈妈，负责引导大家前进的方向，天鹅爸爸断后，负责警戒护卫。

　　A grey heron (*Ardea cinerea*) is standing in the lush reeds and preying onthe unsuspecting fishes in the water. A family of mute swans happens to pass by the heron with their six babies. Swimming at the very front of the fleet is Swan Mom, who serves as the steer, whereas Swan Dad waits on at the end, responsible for the security of the whole family.

◔ 守护·内蒙古乌梁素海
On Guard • Ulansuhai Nur, Inner Mongolia

　　虽然我已经很小心地慢慢地靠近这个疣鼻天鹅的巢区，但是它们早就感知到了。天鹅妈妈带着刚出生不久的宝宝们躲开了，天鹅爸爸则威武挺身，准备阻挡入侵者，护卫自己的家园。

　　Careful as I had tried to be in approaching this nest-dense zone inhabited by the mute swans, they had already long registered me on their vigilant radar. Whereas Swan Mom had dodged away with their young children to places of safety, Swan Dad was plucking up his courage and ready to stand up against me — the intruder to their homeland.

羽翼渐丰·新疆巴音布鲁克
Fledging Swans • Bayanbulak, Xinjiang

　　新疆巴音布鲁克大草原，是一大片广袤的高原湿地，是大天鹅的繁殖地，是国家级的天鹅自然保护区。到了秋季，当年春天出生的大天鹅幼雏已经羽翼渐丰。它们经常扇动翅膀练习臂力，为即将到来的远行做好准备。

　　The Bayanbulak Grassland located in Xinjiang is a vast highland wetland. It is not only a breeding ground for whooper swans, but also a National Nature Reserve for the bird. By autumn, young whooper swans born in spring would have been fully fledged. They would often flap their wings to build up their muscles in preparation for their upcoming long migration.

风雪中的一家·新疆库尔勒
A Family Caught in the Storm • Korla, Xinjiang

两只成年的大天鹅带着五只天鹅幼崽在冰面上溜达。虽然是风雪交加，但它们走得依然轻松愉快，没有丝毫的畏惧。这肯定是得益于体表厚实的羽绒保持体温，但它们的脚趾是裸露的，难道就不怕冻伤吗？看完第一章的内容，相信你已经了解了。

Two adult whooper swans are taking the five of their chicks for a stroll on the ice. Despite of the strong wind and the heavy snow, they all appear to be in a fairly spirited state, not troubled or daunted in the least. This can be attributed to the thick plumage of firs that keep them against the freezing temperature. Nevertheless, aren't they afraid of frostbite to their toes that are exposed bare in the air? Well, if you have already read the first chapter, I guess you would probably have already got the answer.

幼鹅·山东荣成
A Whooper Swan Chick • Rongcheng, Shandong

一只当年出生的大天鹅幼鹅正在梳理羽毛。羽色主体为灰色，其中还夹杂一些白色。嘴部黄斑的颜色还不够深。幼鹅要长到三至四岁才能成年。

A whooper swan chick born earlier this year is grooming his feathers, whose predominantly gray plumage is occasionally dotted with white ones. The yellow spots on its mouth are not yet dark enough. Young whooper swans will not reach adulthood until they are three or four years old.

争鸣·青海贵德
Challenging the Authority • Guide County, Qinghai

　　大天鹅一般结伴迁徙。通常是有血缘关系的一起组队。领队是有经验的长者，大家都比较听从他的指挥。但在集体生活中也难免会发生一些争执。有时它们会扇动双翅，抻长脖子，扯着大嗓门发表自己的意见。但嚷嚷过后就会慢慢地平息下来，然后该干嘛就干嘛去了。毕竟是自家人嘛，别伤了和气。

　　Whooper swans usually migrate in companies. In general, flocks of birds in their kinship would assemble into a larger group that is commanded by a senior and more experienced member of the community, whose orders would be duly observed and followed by the entire group. But there might inevitably be some disputes when you live in large groups. In such cases, the rebellious ones would stretch their necks out and flap their wings vigorously, as if to voice their defiance against the authority. But they will soon calm down after the commotion and regain their regular routines. After all, they are all of the same family, and only families living in harmony will prevail.

争吵·山东荣成
A Minor Quibble • Rongcheng, Shandong

　　如果大天鹅的群体数量比较多，密度比较大，彼此争吵的现象就会时有发生。争吵通常发生在两个携带幼鹅的家庭或家族之间，原因无非是彼此距离过近，感觉不自在了。双方都是为了自家的安全和舒适与对方争吵不休，最后通常是靠武力来解决争端。

　　If a densely populated community of whooper swans, minor quibbles would break out frequently. Such unpleasant events typically happen between two families with young chicks that have come too close in the vicinity and feel ill at ease for the presence of the other. Motivated by concerns over the safety and comfort of family members, such disputes would often come to an end by resorting to violence.

◀ 打斗·河南三门峡
A Fierce Battle • Sanmenxia, Henan

发生争执的大天鹅家族，如果双方势均力敌，则打斗的场面会比较激烈，如翻江倒海，但很快就能分出胜负。

通常，参与打架的是两只成年的大天鹅，它们的幼鹅都在一旁观战，呐喊助威。偶尔也会看到集体出动，群殴的场面。

Fights that break out between two swan families that are on relatively equal footing in strength would often be extremely fierce but fast-paced. It won't take long to see which side will prevail in the end.

In general, the ones who join the fights on behalf of the two parties would be the adults of the whooper swan families involved, with their young chicks watching on as the cheer-leaders. However, massive and disorderly group fighting does occur occasionally.

HABITATS OF SWANS
IN CHINA

天鹅圣境

征

　　在中国能见到的野生天鹅，都有每年随着季节变化从繁殖地到越冬地往返迁徙的行为特征。从飞行的距离上看，可以说每次都是长征。繁殖地通常人迹罕至且隐蔽性很强，但迁徙途中停歇地的环境有许多安全隐患及食物供给的不确定性。最重要的是，离开繁殖地的天鹅家庭里有许多当年出生的幼鹅。它们体质较弱、飞行能力不强，漫漫旅途对于这些幼鹅来说是一场生死考验。

　　天鹅的体重较大，起飞时会一边扇翅一边奋力奔跑，在水面上踏起一长串的浪花。在空中翱翔的天鹅姿态十分优美，那长长的脖颈和舒展的双翅简直是美到了极致。要降落时，天鹅的双翅和双脚微垂，借助风的阻力惬意地滑翔。当快要接近水面时，天鹅的双脚向前蹬直，踩水减速，激起浪花一片。观看天鹅的飞翔，就是在享受一场视觉的盛宴。

Almost all swans you see in wild in China have the habit of taking annual long migratory journeys between their breeding grounds and wintering grounds as the seasons evolve. Judging in terms of the distance covered during such journeys, each one can be counted as an odyssey. The places where they breed are usually located at some remote and densely sheltered sites that are not easily accessible by human beings, whereas potential security risks and uncertainties in food supply abound in places where they stop over for rest during their long migratory trips. What's more, there are quite a number of newly born baby swans that tend to be physically weak and vulnerable in long distance flights. For these young swan chicks, such odysseys would often turn out to be matters about life and death.

Swans tend to be comparatively large in size. To take off successfully, they will have to flap their wings swiftly and vigorously as they run ahead at neck-breaking speed, stirring up long waves behind them. Highlighted by their slender necks and out-stretching wings, swans soaring in the high sky are quite elegant and graceful — impressing viewers with a sort of beauty this is just incredible. At times when they are about to land, they will, with their wings and feet slightly drooped, take advantage of the resistance of air-flow to glide across the sky in an elegant manner. As they approach the water surfaces, they will stretch their legs straight forward to slow down through treading on water surface, raising a string of gleaming foams as they splash down. Watching the flying of swans is truly a visual feast.

➲ 起飞·青海青海湖
Taking-off • Qinghai Lake, Qinghai

　　大天鹅在游禽当中属于体形、体重都比较大的鸟类，所以起飞比较费力。通常，它们会选择较宽阔的水面，借着逆向的风，猛扇翅膀奋力奔跑，以便用最省力的方式及最短的距离飞向天空。

　　Among all the natatory birds, whooper swans are of comparatively large size and weight, thus making it more energy consuming for them to take off. They often choose broader water surface as their takeoff ground, where they would run against the airflow and flap their wings vigorously, so as to elevate themselves up into the air in the least energy-consuming manner and within the shortest distances.

➤ 跟拍飞翔·河南三门峡
Snapshot of a Fast-flying Swan · Sanmenxia, Henan

　　大天鹅在飞翔时，无论如何剧烈地扇动翅膀，其头部基本保持稳定。根据这个特点可以进行慢速跟踪拍摄。这样就可以得到头部相对清晰而翅膀虚动的照片。

　　When whooper swans fly in the sky, no matter how vigorously they flap their wings, their heads will remain basically stable. If you know this defining feature of them, you will be able to take advantage of it to capture fairly good pictures of swan on flight, with its head clearly locked down in focus and its wings a bit blurry.

● 集体舞·河南三门峡
Group Danding • Sanmenxia, Henan
● 动感天鹅·河南三门峡
Dynamic Swan • Sanmenxia, Henan

　　"动感天鹅"的拍摄成功率不高，主要是因为相机跟拍的稳定性不好控制，不容易拍实。相机加长焦镜头比较重，需要粗壮的三脚架和阻尼适度的优质云台的辅助支撑，才能拍出虚实得当的效果。

　　It is not easy to succeed in capturing a good photo about "dynamic swan", primarily because of the difficulties in keeping the camera stable enough to take in solid images. In addition, the camera with telephoto lens is rather heavy, which must be backed properly up by sturdy tripods and quality pan-tilt with proper damping to capture pictures that come with clear views.

◐ 来客·新疆伊犁
Faraway Guests • Ili, Xinjiang

每年的冬季，会有一些来自哈萨克斯坦巴尔喀什湖的疣鼻天鹅，到新疆的伊犁河谷湿地越冬。队伍后面的一只天鹅的脖子上的环志标牌，表明了它们的身份。

Each year, flocks of mute swan that have come from the faraway Balkhash Lake in Kazakhstan will arrive in succession at wetlands in Ili River Valley in Xinjiang for wintering. The bird banding on the neck of a swan flying at the rear end of the flock group reveals their identities.

◐ 迁飞·青海青海湖
Migratory Flight • Qinghai Lake, Qinghai

青海湖进入了冬季。随着气温的逐渐降低，湖面结冰的范围也在增加。大天鹅迁飞到泉眼附近有活水的地方，以满足基本的生存需求。

The Qinghai Lake is entering into the winter with temperature gradually getting cold and the range of melted area increasing. Whooper swans begin to migrate to the spring opening to get flowing water for basic survival needs.

◑ 路过·辽宁朝阳
Stopover • Chaoyang, Liaoning

辽宁朝阳市郊的红村附近，每年都会有迁徙的大天鹅从这里路过。它们会在大凌河湿地停歇，有一小部分还会留在这里越冬。

Vast number of migrating whooper swans would pass by the Hongcun Village that sits on the outskirts of Chaoyang, Liaoning Province each year. They would stop over at the Daling River Wetland for a short break before going ahead with their journey, and a small part of the birds would choose to settle down and spend their winter here.

❂ 继续南下·河北沽源
Further Southwards • Guyuan, Hebei

这是深秋的季节，一群小天鹅在沽源的囫囵淖尔做了短暂的停歇后准备继续南下。

It is already in late autumn. A flock of tundra swans, after a short recuperative stop-over in Hulun Nou'er in Guyuan, are getting ready to resume their south-bound long odyssey.

❂ 飞过城市·安徽池州
Passing the City • Chizhou, Anhui

安徽池州位于长江的中下游。秋季会看到一群群小天鹅飞过城市的上空，寻找合适的地方停歇或越冬。

Situated on a favorable site in the middle and lower reaches of the Yangtze River, Chizhou City in Anhui Province would in each autumn witness flocks of tundra swans soar graceful overhead on their odyssey to find ideal niches for either a short stopover or spending their winter times.

早到的客人·北京颐和园
Early-arriving Guests • The Summer Palace, Beijing

每年的初春，颐和园都会迎来一群群迁徙的候鸟在此停歇。小天鹅是它们当中比较早的一批。每当天鹅来临，昆明湖就变成了天鹅湖。

Flocks of migratory birds would splash down in the Summer Palace each year in early spring for short stopovers. The tundra swans are often among the early-arriving ones that turn the famous Kunming Lake in the park in to one more SwanLake favored by tourists.

华丽的舞步·新疆赛里木湖
Gorgeous Dancing • Sayram Lake, Xinjiang

新疆的赛里木湖海拔两千多米，天气变化很快。刚才还是晴空万里，忽然乌云就出现在了天边。这场景就像是舞剧《天鹅湖》的剧场。在幽暗的背景下，一只洁白的大天鹅优雅地跳着华丽的舞步。

The Sayram Lake in Xinjiang, at an altitude of 2000 meters, is widely known for its highly changeable weather. Cloudless and sunny sky could in a matter of just seconds be covered with thick layers of dark clouds. This fast-changing backdrop provides an ideal stage for the Swan Lake, the world-famous dancing opera. A snow-white whooper swan is dancing her graceful dances against the dim-lit backdrop of the sky.

⊙ 准备远征·新疆赛里木湖
Getting Ready for the Migratory Odyssey • Sayram Lake, Xinjiang

气候的变化预示着迁徙季即将到来。几只大天鹅在湖边起起落落做着迁飞的准备。崇山峻岭虽然是前进的障碍，但是它们已经立下了远征的决心。

Climate change predicts the coming of a migration season. Several whooper swans are working on their intensive training course for the upcoming long-distance expedition through repeated drills in takeoff and landing. Although there are thousands of high and steep mountains on their migration journey, they are determined to embark on their journey.

⊙ 迁徙的尾声·内蒙古达里湖
Towards the End of Migratory Season • Darry Lake, Inner Mongolia

深秋的季节，当我站在山顶望向辽阔的达里湖，能依稀地看到一些大天鹅在岸边徘徊。寒冷的北风凛冽刺骨，湖边的水面已经结冰。据当地人讲，这些天寒流来袭，大部分天鹅已经向南飞走了。如果持续降温，湖面的结冰面积增大，剩下的天鹅找不到东西吃就待不了多久。此时已经接近天鹅迁徙的尾声了。

It's late autumn. As I stood at the hill top and cast my eyesight to the broad Darry Lake, I caught sight of a dim view of some whooper swans hanging about the shore. The chilly northern wind was piercingly cold and the water surface had already frozen. According to local people, the sudden arrival of a cold spell in the last few days had sent most swans on their south-bound journeys. If the temperature kept dropping, the few swans that still lingered on would soon find it difficult to get enough food as the lake got increasingly covered up in ice and therefore would have no choice but to fly away. The peak migratory season in the lake was drawing to its close.

◀ 飞行训练·内蒙古达里湖
Flying Drill • The Dalihu Lake, Inner Mongolia

大天鹅的一家六口在即将冰封的湖面上盘旋。它们喜欢这里的安静舒适，喜欢这里的水草丰美。但它们必须做好继续南迁的准备。

A family of 6 whooper swans are hovering around the soon-to-be frozen lake. Much as they are fond of this quiet and tranquil place endowed with copious water and abundant grasses, they will have to get themselves duly prepared for the southward migration that is coming up soon.

Defying the Harsh Weather • Rongcheng, Shandong

　　风云突变，一队大天鹅在逆风向前。它们明白，迁徙的过程不会总是一帆风顺的，有时也需要风雨兼程。

The weather changes suddenly and a flock of whooper swans are flying painfully against the wind. They know perfectly well that the migratory journey will by no means be an easy one and that flying in defiance against the weather would sometimes be inevitable.

▶ 接纳"丑小鸭"·山东荣成
A New Family for the "Ugly Duckling" Astray • Rongcheng, Shandong

　　通常，天鹅的集群都是同一种类。即便有不同种的天鹅同时在一个地区出现，它们彼此也保持一定的距离。但也有个别失群的孤儿被其他种类的天鹅家族收留的实例。我曾见到过单只疣鼻天鹅混到大天鹅的家族中共同生活。更神奇的是，我还见到过一只鸿雁混入了大天鹅的家庭，成为家庭成员，与大天鹅一起生活，一起编队飞行，这简直太罕见了。

　　我推测，那只鸿雁一定是在迁徙途中掉队了，成了孤雁。求生的本能使它不断地寻找同行的伙伴。天鹅与鸿雁同属游禽，习性相似，迁徙途中的停歇地基本重合。但天鹅生性孤傲不好相处，对于外来者都很排斥。而这个天鹅家族却触动了善心，接纳了这只"丑小鸭"。可见生物的多样性与复杂性。

In general, only swans of the same species would live in the same community. Even if swans of different species do show up simultaneously at the same site, they will somehow manage to keep a distance in between. However, there are also cases when an individual bird of a different species that has gone astray is adopted and accepted by the family of another swan. I have personally seen a single mute swan living in perfect harmony with a whooper swan family. More surprisingly, I have even seen a swan goose (*Anser cygnoid*) that has made its way into the family of whooper swans and become one of its members. The goose lives and flies together with the family, which is really rare to see.

I guess that particular swan goose must have fallen behind on its migratory journey and got astray. Driven by an instinctive desire for survival, it has to find new companions for the journey ahead. Swans and swan geese all belong to natatory birds that share roughly similar life behaviors and migratory routes. However, given that swans are often quite proud birds that tend to be highly wary of foreign intruders and therefore difficult to get along with, it is all the more amazing and unusually that this particular family of swans has been kind enough to have accepted the "ugly duckling" as one of its members. What a convincing evidence for the diverse and intriguing nature of the ecosystems of the Earth!

◀ 仰视·山东荣成
An Upward Snapshot • Rongcheng, Shandong

这只大天鹅忽然迎面飞来，几乎擦着我的头皮飞过。还好我的相机不离手，抓拍到了如此仰视的近身照。这完美的身姿真是令人赞叹。

This whooper swan flew right up to me unexpectedly, just above my head. Fortunately, I have the habit of keeping the camera ever handy, so I am able to take such a close-up snapshot of the whooper swan in a upwards looking position. Its near perfect figure is just amazing.

131

❯ 检阅·山东荣成
Inspection • Rongcheng, Shandong

一队大天鹅像一群轰炸机，排着整齐的编队从我的头顶附近掠过。让我可以用这样的视角"检阅"它们的美。

Like an escadrille of bombers, the whooper swans flew in neat formation over my head, allowing me a precious chance to "inspect" their beauty from such a perspective.

迁徙途中·辽宁朝阳
On the Way • Chaoyang, Liaoning

朝阳的白石水库周边有一大片湿地浅滩，非常适合天鹅等游禽栖息，已然成为大天鹅在迁徙路途上的加油站和避风港。

There is a vast stretch of shoal wetland around the Baishi Reservoir in Chaoyang, which has long become a most-favored top-up station where natatorial birds like whooper would stopover for food and rest during their journeys.

降落水面的瞬间·山东荣成
Splashing Down at the Lake • Rongcheng, Shandong

这张照片近距离地记录到了大天鹅即将降落到水面的瞬间。只有经过观察，预判天鹅有可能降落的地点并提前到附近守候，才有可能拍到这样的画面。

Captured in this photo is the moment when a group of whooper swans are about to splash down into the lake. Such a close-up snapshot can only be capture when you, on basis of careful observation, can accurately anticipate the exact site where swans are most likely to land and manage to get there early enough and wait patiently.

137

> 秋韵•河南省三门峡
Autumn Rhythm • Sanmenxia, Henan

　　十月下旬，秋染三门峡的黄河湿地。一批批远道而来的"客人"——大天鹅，陆陆续续来到这里，使这里的秋景别有一番韵味。

It's late October when the Yellow River wetland at Sanmenxia has been tinted with the colors of autumn. Flock upon flock of groups of "guests" — whooper swans — are gradually arriving from afar to spend their winterhere. Their arrival adds an extra tinge of romance and pleasantness to the autumn scenery here.

纷至沓来·山东荣成
Arrival in Succession • Rongcheng, Shandong

日暮时分，一群大天鹅纷至沓来，准备降落休息。这使得原本宁静的港湾一下子热闹了起来。

夕阳虽红，但穿过淡淡的雾霭已经没有了刺眼的光。就像这些归来的大天鹅穿越了万水千山，看上去已然是身心疲惫。荣成的滨海湿地就是它们理想的度假胜地。

As the sun is about to set down, a group of whooper swans are arriving in succession and getting ready for spending their nights here, which makes the quiet harbor suddenly filled with the vibrancy of lives.

Red as the setting sun still is, the gathering mists have already screened its dazzling light away, just like these whooper swans that are returning from their long and arduous journeys, worn totally out both physically and mentally. The coastal wetlands in Rongcheng afford them an ideal resort for they to recover from their long and tiresome migration.

栖

对于天鹅来讲，繁殖地就相当于它们的家，迁徙的停歇地就相当于旅馆，而越冬地就算是度假村了。笼统来讲，这些都是天鹅的栖息地，共同特点就是适合水禽生活的湿地，并且这些湿地的类型也是多样的。

通过实地考察发现，天鹅夏季的繁殖地多在水草茂密的湖泊沼泽湿地，这类湿地的特点是隐蔽性好、安全性强，不易被发现、被打扰，并且食物充足。春秋的迁徙停歇地类型就多了，只要是相对安全，能解决基本食物问题，就能停歇。越冬地除了安全，对于食物的要求要高一些。因为是长期驻留，食物的量必须要充足。

越冬地的天鹅其实就是在度假。它们在养精蓄锐为回迁作准备，一些适龄的"年轻人"也在寻找伴侣、谈情说爱。虽然回到繁殖地的路途依旧漫长，但那里是它们的家。它们出生在那里，它们的后代也将要在那里出生。

For swans, the breeding grounds are just like their homes, the places where they stop over for rest and energy top-up are like their hotels, and the habitats where they spend their winters can be compared to their holiday resorts. Generally speaking, all these places are swan's habitats that have a common feature — all being wetlands that are suitable for waterfowls to live, and falling into different types of wetlands.

Field observations I personally made conclude that the breeding grounds of swans in the summer are mainly distributed in lakes and swamps with dense water grass. Such wetlands tend to be better covered up and therefore afford better security. They are not easy to be found or disturbed and they have abundant food. As to the places where they merely stop over for short breaks in spring and autumn, there are obviously many more options, so long as sufficient amount of food is available. In addition to providing safe and secure shelters, food supply in their wintering places must be comparatively more abundant, given that the time span they spend there would be much longer.

In fact, swans in wintering ground are just on their vacation and preparing for moving back. Some young swans of marriageable age also find their suitable mates during this period. Although there is still a long distance to go before they make their ways back to their breeding ground, it is their home after all. They were born there, and so will be their descendants.

⊙ 冬日里的生机·辽宁朝阳
Vitality in Winter • Chaoyang, Liaoning

大凌河流经的朝阳红村附近，地势平缓。在冬季的枯水期，河道会呈现浅滩。河面虽然结冰但不会全部封冻，因此河道附近的空气湿度大，有时会出现雾凇的景观。大天鹅的降临，给这严寒的冬日平添了许多生机。

The Daling River flows near the Chaoyang Village where terrain is flat. During the dry season of winter, the river channel will show its shallow shoal. Though there is ice on the surface, the river will not be frozen entirely because of the high air humidity and sometimes there is rime. The appearance of whooper swan brings many vitalities to these cold winter days.

⊙ 晨曦·辽宁朝阳
First Ray of Sunlight • Chaoyang, Liaoning

日出东方，薄雾渐散。睡醒的大天鹅正在"梳洗打扮"，准备迎接崭新的一天。

The sun rises in the east, and the mist dissipates. The whooper swan who wakes up is "dressing up" and getting ready for a new day.

◀ 暮归·内蒙古乌梁素海
Returning at Dusk • Ulansuhai Nur, Inner Mongolia

夏季的傍晚，水面映衬着五彩云霞，疣鼻天鹅一家正在归巢。优良的生态环境使这里成为国内最大的疣鼻天鹅繁殖地。

On a summer evening, with colorful clouds reflected on the water, a family of mute swans are returning to their nests. The excellent ecological environment makes it the largest breeding ground for mute swans in China.

▶ 锦上添花·青海贵德
The Extra Touch of Majesticness on the Crown • Guide, Qinghai

黄河在青海贵德这一段水位不深，水质清澈。两岸的植被以多枝柽柳为主，山体为多彩的丹霞地貌，风景本来就十分奇美。大天鹅来这里越冬，为这里的美景锦上添花。

The Yellow River at the Guide section in Qinghai is not deep, with crystal-clear water. The vegetation on both sides is dominated by tamarisk (*Tamarix ramosissima*), and the mountains are of the colorful Danxia landforms. The scenery is very beautiful. Whooper swans come here to spend the winter, adding a further trimming to the wonderful views typically found here.

● 短暂的停歇·北京颐和园

A Short Stopover • the Summer Palace, Beijing

　　初春的颐和园昆明湖，水面浮冰形成的"小岛"成为小天鹅停歇的落脚点。它们在此做短暂休整后将继续北上。

　　In the Kunming Lake of the Summer Palace in early spring, "the small island" formed by the floating ice on the water has become a favorite stopover site for tundra swans. They will continue flying northwards after a short break here.

◀ 小憩·河北沽源
A Relaxing Break • Guyuan, Hebei

　　深秋，河北沽源的囫囵淖尔已经开始结冰。南迁的小天鹅在这里小憩，缓解旅途的疲劳。

　　In late autumn, the Hulun Nou'er (the Swan Lake in Manchu language) in Guyuan, Hebei begins to freeze. The south-bound tundra swans are taking a rest here to relieve them of the fatigue of the long journey.

享受 • 山东荣成
Indulgence • Rongcheng, Shandong

夕阳西下，和风徐徐。三只大天鹅都眯着眼睛享受着美好的时光。

The sun is setting, and the wind is gentle. Three whooper swans are all squinting their eyes and enjoying the good times.

睡觉 • 山东荣成
Sleep • Rongcheng, Shandong

天鹅睡觉时会将脖子扭到背部，把嘴插到翅膀里面。但是眼睛会露在外面，一旦感觉到动静就会睁眼观察。

When swans sleep, they would twist their necks over their backs and tuck their mouths beneath the wings. But the eyes will be exposed, and once they feel any slight disturbance, they will open their eyes vigilantly to observe.

163

◆ 涂抹·山东荣成
Skincare Time • Rongcheng, Shandong

　　大天鹅尾巴的油脂腺会经常分泌出油脂。它们会将嘴伸到尾巴附近抹擦，一边梳理羽毛，一边将油脂涂抹到全身。这样就保证了羽毛不会被水浸湿。

　　The oil glands at the tail of whooper swans often secrete oil. They rub their mouths near their tails, combing their feathers while applying the grease all over their bodies. This ensures that the feathers would not be soaked as they dip into water.

➤ 梳理羽毛·山东荣成
Grooming • Rongcheng, Shandong

有时大天鹅梳理羽毛会很仔细。里里外外，上上下下，几乎每一根都能照顾到。

Sometimes whooper swans will comb the feathers very carefully, inside and out, up and down, leaving not part of the body uncared.

◐ 洗澡·山东荣成
Bath-taking • Rongcheng, Shandong

　　大天鹅虽然几乎天天都泡在水里，但并不是可以不洗澡了。它们时常会拍打水面将全身的羽毛打湿，痛痛快快地洗个澡。

　　Although whooper swans spend most of their times in water, it doesn't mean that they don't need to take bathes. They would often splash the surface of the water to wet the feathers all over their bodies so as to take a good bath.

169

抖掉水滴・山东荣成
Shaking off Water Droplets • Rongcheng, Shandong

大天鹅在喝水或者洗脸后，会抻着脖子快速地抖掉羽毛上的水分，以保持身体的干燥。

After drinking or washing their faces, whooper swans would quickly stretch their necks out and try to shake off water drops accumulated over their feathers so as to keep themselves dry.

◀ 踏浪·山东荣成
Riding the Waves • Rongcheng, Shandong

　　起风了，海水荡漾。大天鹅随着波浪起伏，似乎很是享受这快乐的时光。

　　The wind is blowing and the sea is rippling. Whooper swans ride over the waves and seem to enjoy themselves quite a lot.

173

➤ 戏水·山东荣成
Gaming the Water • Rongcheng, Shandong

大天鹅不断地将脖子探入水中又提起，激起一片水花。这是在梳洗，也是在玩耍。

The whooper swan keeps probing its neck into the water and lifting it again, splashing up white foam around it. This is their way of grooming, but also a way for entertaining themselves.

◀ 振翅•山东荣成

Flying High • Rongcheng, Shandong

一只强壮的大天鹅抖擞精神起身振翅，一派王者风范。周围的天鹅纷纷低头，似乎在表达对它的臣服。

A strong and stout whooper swan stirs his plumes and takes off to the air in majestic manners. All other swans around it bow their heads hurriedly down, as if paying their homage to their royal king.

● 初醒·河南三门峡
Waking Up • Sanmenxia, Henan

伴随着红日初升，大天鹅与城市一同醒来。

As the sun rises, the whooper swans are also waking up along with their human neighbors in the city.

● 共生共存·河南三门峡
Harmonious Coexistence • Sanmenxia, Henan

逆光下水天一色。都市的楼群、桥梁与水面上的大天鹅都成为剪影。这凝练的画面突显了建筑物的庞大与天鹅的渺小。巨大的反差让我开始思考：高速发展的人类社会与大自然应该如何和谐相处？如何共生共存？

The water and the sky appear to be of one color under the backlight. The urban buildings, bridges and the whooper swans are all silhouetted. This condensed image highlights the massiveness of the buildings and the smallness of the swans. The sharp contrast between the two throws me into thinking: how could human society and nature exist side by side in today's face-paced world? what are the ways for this harmonious coexistence?

● 晨曲·河南三门峡

Morning Song • Sanmenxia, Henan

黎明时分，大天鹅们有的还在蒙头大睡，有的在梳洗打扮，有的已经出发觅食。它们如音符一般或慵懒闲适，或明快活泼。

At dawn, some whooper swans are still sleeping while some others are already grooming themselves up or on the way for foraging. Just like the different notes in a musical score, some of them are spirited and uplifting, whereas some others appear to be more tranquil and slow-paced.

183

◖相映成趣·河南三门峡
A Delightful Contrast • Sanmenxia, Henan

前景的一排残荷的剪影与远景的一排大天鹅形成实与虚的对比，彼此相映成趣。

The silhouette of the row of fading lotus in the foreground forms a pleasant contrast with the rank of whooper swans that happens to be flying by in the distance.

◖包围·河南三门峡
Entrapped • Sanmenxia, Henan

一群白骨顶一边觅食一边游荡，渐渐地将一只不知所措的大天鹅围住。这画面很有几分喜感。

A flock of common coots (*Fulica atra*) roam aimlessly as they forage, unwittingly getting a bewildered whooper swan entrapped among them. There is a kind of indescribable sense of comic touches in the photo.

◗雪中乐·河南三门峡
Merry-making in the Snow • Sanmenxia, Henan

大雪纷飞的日子，大天鹅通常都在原地休息，只有少数的"活跃派"不时地飞来飞去。即便是能见度不高，也不能阻碍它们的飞行。

Whooper swans usually would stay quietly in the lake on snowy days, with the exception of a few "hyper-active" ones who would defy the poor weather and the low visibility to flutter around now and then.

散步·山西平陆
A Walk on the Land • Pinglu, Shanxi

尽管大天鹅喜欢在水里待着，但时常也会到岸边的草地上走走，摇摇晃晃的样子像只笨鸭子。

Although whooper swans tend to spend much of their time in water, they occasionally would also like to take short walks on the grasslands. The way they walk unsteadily makes them look like clumsy ducks.

◀ 静谧·河南三门峡
Serenity • Sanmenxia, Henan

在三门峡市郊的王官村附近，有一处较高的位置可以俯瞰一部分黄河湿地。越过近前的树梢，可以看到静谧的水面上飘荡着的大天鹅。

Near Wangguan Village on the outskirts of Sanmenxia, there is a high spot that commands a panoramic view of a certain section of the Yellow River wetlands. Casting your eyesight over the treetops in front of you, a group of whooper swans swimming leisurely in the wave-less lake would come into your sight.

◀ "破冰行动"·河南三门峡
"Operation Icebreaker" • Sanmenxia, Henan

降温了，湖面上结了一层薄冰。几只大天鹅用嘴掰断冰碴，向前开路。这样的现象还是比较罕见的。

The temperature has dropped and a thin layer of ice has formed on the surface of the lake. A few whooper swans broke the ice with their beaks and made their painful ways ahead. This is indeed a rare view to come across.

▶ 金秋时节·河南三门峡
Golden Autumn Season • Sanmenxia, Henan

金秋的季节，大天鹅一家早早就来到了越冬地，享受着明媚的阳光和安宁的生活。

In the golden autumn season, the whooper swan family arrived as the early-birds at this favorable wintering place where bright sunshine and peaceful life can be expected.

徜徉·山西平陆
Roaming in Leisure • Pinglu, Shanxi

　　山西平陆的三湾紧靠黄河北岸，与河南的三门峡隔河相望。这里也有一片宽广的湿地，同样也是大天鹅的越冬地。附近的山上有一个"制高点"，是俯瞰天鹅湖的理想之地。

　　Sanwan in Pinglu, Shanxi Province sits on the northern bank of the Yellow River, just opposite to Sanmenxia in Henan Province that sits on the other side. There is also a wide stretch of wetland here where whooper swans often spend their winters. The top of a hill nearby commands an advantageous position for watching the swans in the lake.

199

❯ 光斑之彩•新疆伊犁
Glittering Necklaces • Ili, Xinjiang

　　我时常会对逆光下的疣鼻天鹅着迷。耀眼的光斑围绕在天鹅的脖子周围，像一串串珠宝，光彩夺目。

I am often intoxicated by the view of mute swans in backlight. Dazzling rings of sunlight would appear around the neck of the birds, glittering gloriously like strings of jewelries.

◐ 迎接日出·新疆伊犁
Ushering in the Sunrise • Ili, Xinjiang

　　红日东升，驱散了晨雾。一只疣鼻天鹅起身振翅，像是在拥抱灿烂的阳光，拥抱美好的未来。

　　The red sun rises in the east, dispersing the morning fog. A mute swan takes on its wings and flies into the sky, as if embracing the brilliant sunlight and a bright future.

◔ 凌波仙子·新疆伊犁
Water-borne Fairy • Ili, Xinjiang

◔ 天仙下凡·新疆伊犁
**The Celestial Fairies That Have Come Down to
This Earthly World • Ili, Xinjiang**

　　见到此情此景，我只能用"天仙下凡"来形容了。新疆伊犁的天鹅泉湿地公园，是近些年喜欢拍疣鼻天鹅的摄影师所聚焦的热门地点。每年冬季，如果遇上雾凇的天气，这里就变成了仙境。

　　The only description that occurred to me when I caught sight of this view was that "they must be the celestial fairies that have come down to this earthly world". The Swan Spring Wetland Park in Ili, Xinjiang has in recent years become a popular spot for photographers to take pictures of mute swans. When rime appears here in winter, this place would turn into a fairyland.

◔ 梦幻之境·新疆伊犁
Land of Dreams • Ili, Xinjiang

　　天气越寒冷，早晨的雾气就有可能越大。疣鼻天鹅飘荡在湖面上，若隐若现，十分梦幻。

　　The colder the weather, the mistier the mornings tend to be. Mute swans floating on the surface of the lake loom behind the misty air, visible for a while and then disappearing suddenly. It feels as if you are in a dreamland.

渐入佳境·新疆伊犁
Closer to Nirvana • Ili, Xinjiang

晨雾笼罩着寂静的芦苇荡，几只疣鼻天鹅时隐时现。这样的景象真是有些妙不可言。只感觉自己的心情无限舒展，随着天鹅的缓慢游动渐入佳境。

A couple of mute swans appeared and disappeared unexpectedly among the mist-enshrouded reeds. There was something just indescribable about such a view, leading me unconsciously into a wonderland where my mind soared to the endless sky as I looked at the graceful shades of the swans swimming in the lake.

后 记
EPILOGUE

这本画册的出版有许多机缘巧合。

首先是《关于特别是作为水禽栖息地的国际重要湿地公约》（简称《湿地公约》）第十四届缔约方大会将于2022年11月5日至13日在中国武汉和瑞士日内瓦同时召开。今年是我国加入《湿地公约》的第30年，并且是首次作为东道主承办这个会议。为了宣传湿地文化，作为对大会的献礼，中国林业出版社自主策划了一系列湿地主题的图书，其中包括《湿地光影丛书》。

其次，我对国内天鹅栖息地的影像考察项目已完成，图片与文字资料齐备，内容符合《湿地光影丛书》的要求，所以中国林业出版社决定予以出版。

在此，要感谢中国林业出版社党委书记成吉对本书的认可与肯定，感谢自然保护分社（国家公园分社）社长肖静的鼎力举荐，感谢特约编辑田红对这本画册的精心策划及专业指导，感谢责任编辑袁丽莉为之的辛勤付出。

回想过去，我能在生态摄影方面取得一些成绩，还要感谢那些曾经帮助过我的良师益友。

我与黄华强先生相识多年，他是我艺术设计方向上的领路人。他本人的书籍装帧作品及邮票设计作品相当精彩。其深厚的艺术造诣令我永远敬仰。有关天鹅专题摄影的早期构想就曾得到他的支持与鼓励，期间也多次请教并听取他的建议。

　　在此还要感谢一下我的中学老师王绍璋。当年，在我比较迷茫的时候，是她帮助我明确了学业方向。现在我从事的专业工作离不开她当年的指导与帮助。

　　在天鹅摄影方面，我要感谢河南省三门峡市的曹晓春老师。她的慢速摄影技术简直出神入化。受她的启发和指点，我的摄影作品也有了一定的进步。

　　还要感谢影像生物多样性调查所（IBE）团队的同仁。感谢曾经的队长徐健，在天鹅专题方面给予的指导和帮助；感谢队长郭亮给予的鼓励和支持；感谢王斌和郑运祥给予的鸟类学方面的指导；感谢曾经一起完成拍摄项目的队员。从你们身上，我不但学到了许多生态摄影的知识和技能，还学到了了不畏艰辛的工作精神。能与你们共事，是我的荣幸。

崔林

2022年9月

The publication of this eco-album is the result of quite a few "lucky strokes" that happen to coincide.

First and foremost, the 14th Conference of Parties to the *Convention on Wetlands of International Importance Especially as Waterfowl Habitat* (known more commonly as *the Convention on Wetlands*, or the *Ramsar Convention*) is soon to be convened on November 5th to 13th simultaneously in Wuhan, China and Geneva, Switzerland. This year marks the 30th anniversary since China signed up to the *Convention on Wetlands* as well as the first time for it to host such a grand event. To promote knowledge about wetlands, and to celebrate the opening of this epoch-making conference, the China Forestry Publishing House takes initiative to publish a series of books that theme on wetlands, including the *Collection of Wetlands in Images*.

Secondly, my IBE expedition to the swan habitats in China has come to a successful conclusion, and I have at my hand a whole set of text and image documents in this aspect. Since all these materials fit in well with the ideas behind the *Collection of Wetlands in Images*, China Forestry Publishing House is, to my great delight, interested to have it included.

My heartfelt gratitude goes to Mr. Cheng Ji, the CPC secretary of China Forestry Publishing House, for his encouraging comments on and unreserved supports to this book; to Ms. Xiao Jing, head of the Nature Conservation Division (the National Parks Division) under the Publishing House, for her supportive recommendation; to Ms. Tian Hong, the guest editor of this eco-album, for her innovative and professional advices on the making of the book; and to Ms. Yuan Lili, in-charge-editor, for her devoted work in making this book ready for publication.

I owe what I have achieved in eco-photographing over the past years to all the supportive mentors and caring friends that have offered me their unselfish helps.

I'd like to extend my thanks to Mr. Huang Huaqiang, a long-standing friend and art mentor of mine, whose artistic tastes and accomplishments in book decoration and post stamp design is something that I admire most strongly. He gave me some highly constructive suggestions in the early stage when my plan for a swan-theming eco-album was still under conception, not to mention the many encouraging and enlightening ideas he provided as the project was put under way.

I also would like to extend my thanks to Ms. Wang Shaozhang, my middle school teacher, who helped me to set up a clear goal for my academic career. But for her guidance and help, I probably wouldn't have embarked on a journey that brought me to my current profession.

In the aspect of swan photographing, I would like to extend my thanks to Ms. Cao Xiaochun from Sanmenxia City, Henan Province, whose techniques for slow-movement shooting are just amazing. Under her patient nudging and encouragement, I have also made some steady progresses in my expertise.

Lastly, my most sincere thanks go to all my IBE teammates. Thank-you to Xu Jian and Guo Liang, the former and current team leader, for their guidance, support and help throughout the project; thank-you to Wang Bin and Zheng Yunxiang for their professional consultations in the field of ornithology; thank-you to each and every team member that I have met during the photographing expeditions, from whom I have not only benefited in terms of techniques and expertise for eco-photographing, but also in terms of unyielding commitments to jobs and responsibilities. It has been andis a great honor to have had the opportunity to work with you.

Cui Lin
September, 2022